Global Citizenship in Nursing

By: Shreen Gaber

Global Citizenship in Nursing

Shreen Gaber
RN,PhD, MScN, BSc
Nursing Administration Department
Faculty of Nursing
Cairo University

STANFORD UNIVERSITY PRESS

2017

First Printing: 2017

ISBN: 978-1-387-46364-0

Stanford University Press

Stanford University Press
500 Broadway St.
Redwood City, CA 94063-3199
California - USA
Phone: (650) 723-9434
Fax: (650) 725-3457
Stanford Press @stanfordpress

Stanford University Press, USA distributes to most countries outside of North and South America. Special discounts are available on quantity purchases by corporations, associations, educators, and others. For details, contact the publisher at the above listed address.

U.S. trade bookstores and wholesalers: Please contact Shreen Gaber Tel: (+20)100-8144971; Fax: (+20) 23657190 or email: Sameh17@cu.edu.eg

Shreen Gaber

To my lovely huspend and children
Thank you. Without your support and persistence, I would have
never accomplished this work.

Content

Acknowledgements

I would like to express my great thanks and appreciation to my family, parents, colleagues, teachers and my students who are always willing to provide their support and guidance. I also appreciate the efforts of (Stanford University Press) acquisitions editors and there guidance, and finally I will never forget my supportive inspiring author "Dr. Sameh Elhabashy".

Shreen Gaber

Preface

This book depicts an attention to the approaches of nursing citizenship as one of the new trends in nursing as a profession and describes advantages, consequences & how to achieve. In addition, to provide the suggested professionalism guidelines, innovative nurse & social responsibilities toward nursing citizenship, illustrate how we can create a sustainable global nurse workforce. The researcher strives to write the book in comprehensive, concise and feasible sequential steps to be easily understood aspiring at help the professional nurses to apply the mentioned approaches and improve the image of the profession. Whether you are a beginner or an experienced health care provider, I hope that you find this book hypothetically enjoyable

Global Citizenship in Nursing

Introduction:

Citizenship& Globalization implies an ethical and moral obligation for professional nurses to enter and function in a worldwide community. Nurses have a long history of reaching out to the "other," providing care for the most marginalized in society, and acting on issues of social justice both locally and globally. This history provides a rich example of nurses as global citizens and a solid foundation for nurses to take up the current complex challenges in relation to global health. Despite a long tradition of the involvement of nurses in the advancement of social justice issues, the integration of knowledge related to social justice into the education of undergraduate nurses and the practice of professional nurses is inconsistent in the North American context. Bekemeier and Butterfield, following a critical review of 3 key documents of the American Nurses Association, concluded that the conceptualization of social justice inconsistent, ambiguous, and superficial (Coss, Lillian, 1989).

Reimer Kirkham and Browne concur that nursing discourse is rooted in individualism, "making sustained collective address of systemic injustices such as poverty, homelessness, stigma, and socialization largely outside the purview of nursing action. Determinants of health such as poverty and homelessness are typically an integral part of community health nursing curricula; however, students often do not have the opportunity for a meaningful clinical practicum with socially disadvantaged populations (Duncan, Leipert, Mill 1999).

Universities, Institutions as an Interactive Community:

Through the latter part of the 20th century, the focus in higher education was on the internationalization" of curriculum, including the provision of international opportunities for students, and participation in international development projects. As we move into the 21st century, incorporating an international dimension continues to be advocated, however, the operationalization of this concept has become increasingly complex. Over the past decade, there has been a shift in focus away from the concept of internationalization, toward the broader concepts of global citizenship, global social responsibility, or social justice within higher education generally, and nursing education specifically. The need for nursing programs to acknowledge the shared responsibilities for examining global challenges in a dynamic world is now part of the reality. For example, global citizenship, environmental sustainability, and social justice are values underpinning the University plan for international engagement(Reimer, Kirkham, Browne, 2006).

Some of the key elements of the plan are the expansion of the international dimension of students' learning and research experiences; the introduction of global citizenship and global perspectives into the curriculum; and more opportunities for study abroad, internships, and community service learning experiences. Congruent with the university's goal related to global citizenship, the Faculty of Nursing's mission includes the statement that "through our leadership in teaching, research, and practice the

3

faculty strives to contribute to the attainment of health equity in the global community" Similarly, at the University of Cairo, the Faculty of Nursing created an International Office to respond to students' increasing interest in global health issues and the rise of requests from foreign institutions and students. The current goal is one of enhancing global reach and building a reputation through the development of international collaborations and networks among nurse scholars; in order to achieve such a goal, fostering the ethos of global citizenship among students and faculty has become an institutional priority (Smith, 2002).

Professionalism in Nursing and Citizenship:

The need for nursing programs to acknowledge the shared responsibilities for examining global challenges in a dynamic world is now part of our reality in higher education. In this book, we argue that professional nurses must not only participate in examining global challenges but must also contribute to the development and implementation of solutions to these challenges. Kelley and her colleagues suggest that social responsibility is strongly linked to the values of professions in general and nursing in particular" and that "the underlying constructs of social responsibility are woven into the fabric of nursing's history and its code of ethics." (Kelley, Connor, Kun & Salmon, 2008).

Indeed, several authors have traced the roots of social responsibility in nursing to the work of nursing pioneers including Florence Nightingale, Margaret Sanger, Lavinia Lloyd Dock, and Lillian Wald. For example, Lillian Wald highlighted the role of nurses in the advancement of justice beyond the individual: "But the old social theory was established in the belief that the individual was supreme; and then, with civilization's advance, responsibility was extended to cover the family with the tribal group. Smith explored the life work of Lavinia Loyd Dock relative to nursing and caring as social responsibility and the implications of this conceptualization for social equality. The author suggested thinking for the development of a "new ideal" of society and democracy. As a result, on the basis of Dock's epistemology, Smith proposed is conceptualization of nursing academic curriculum "as democratic and as embracing caring as social responsibility for the holistic welfare of others." (Bekemeier, Butterfield 2005).

Concepts are Related to Global Citizenship:

Global citizenship as a concept has its historical roots in the late 20th-century push to internationalize higher education in response to increasing globalization. There are varied conceptualizations of globalization. For example, Hirschfeld describes globalization as "... the worldwide integration of economic and financial sectors, which was made possible by three crucial developments: technical progress, geopolitical changes, and the dominant ideology of regulation of the market.Chen and Berlinguer, on the other hand, argue that a broader definition of globalization is required to capture the complex interplay of the economic, political, cultural, and social dimensions of globalization. With this increasing interconnectedness, globalization can be viewed as unifying and merging the world as being together. Conversely, globalization may ignore the importance of social and human capital with the well-being of people ranked lower in relation to the economic interests (Knight, 2008)

Likewise, Hanson describes globalization as a multifaceted phenomenon with one of its major dimensions being the internationalization of education. Although Torres and Rhodes are contradicted to define higher education impact. All of these authors view globalization as one of the key influences on the process of internationalizing higher education in nursing. Allen and Ogilvie suggest that internationalization at universities is occurring within the context of globalization. Their vision is congruent with Knight's definition of the internationalization of universities as "the process of integrating an international, intercultural, or global dimension into the purpose, functions, or delivery of the nursing higher education. She argues that although the international dimension of higher education is increasingly viewed as important, it concurrently is becoming more complex. Developing one uniform definition for internationalization diminishes the opportunity for contextualization of activities to ensure their relevance within various sectors and countries around the world.

Knight differentiates between internationalization "at home" (faculty -based activities) and "cross border" (Acad, 2010).

In keeping with current thinking within many academic institutions relative to the importance of "internationalization," Knight strongly advocates that as "internationalization matures" one must critically and closely monitor the trends in order to track and respond to short term and future intended and unintended consequences.

Social Responsibility and Citizenship:

Social responsibility, as a concept, has been more critically and broadly discussed in disciplines outside of nursing. In the private institutions and organizations, for example, it is referred to as *integrate social responsibility* which meaning that corporations have moral obligations as "rules and deities of good citizenship" separate from those determined by law. This concept has also been widely studied with in nursing, in response to health care shifts from a "hospital- based orientation" to a "community-centered health promotion focus". Mayo stressed the need for nurse educators to examine more closely the concept of social responsibility (Chavez, Peter, Gastaldo 2010).

Nurse educators were just beginning to examine the social context and processes involved in the students' development of knowledge, values, and practice methods related to social responsibility. Mayo emphasized that nurses working in community-based practices must possess elements of cultural sensitivity, moral and professional practice obligations, and other aspects of social responsibility (Association of Universities and Colleges of Canada, 2004).

The definition and the concept operationalization of social responsibility are increasingly being discussed in the nursing literature. The importance of the concept in relation to global citizenship and social justice has been highlighted. Global citizenship has been discussed across disciplines including education, law, community health and epidemiology, and nursing. although there is no a unified definition for the term" global citizenship," a common thread is the notion of interconnectedness. Citizen responsibility goes beyond the local and national arena, given that some actions impact the planet as a whole, thus creating demand for global responsibility. Byers provides one possible definition for global citizenship: "Global citizenship empowers individual human beings to participate in decisions concerning their lives, including the political, economic, social, cultural, and environmental conditions in which they live. It is expressed through engagement in the various communities of which the individual is part, at the local, national, and global level (Chavez, Peter, Gastaldo. 2010).

Keeping and Shapiro argue that if one agrees that global citizens have ethical obligations to communities beyond their borders, after that what definitely are those obligations? Although the concept of global citizenship still lacks clarity, there is a movement advocating the ethical obligations of all citizens. In a globalized world with moral or ethical responsibilities, nursing students and faculty are increasingly challenged to be competent practitioners. One of the significant challenges is to incorporate the concept of global citizenship to align with nursing's obligations of social justice for all. In the last decades, global health and global citizenship have a rarely been recognized as important concepts to be incorporated into nursing curricula.

Crigger and colleagues articulate a global ethic for nursing, emphasizing the concepts of "world citizenship" and "compassionate professionals" as important components to include in the education of students and faculty in support of a commitment to social justice and global health. In reviewing the current literature, it is evident that global citizenship and social responsibility are increasingly important concepts relative to the interconnectedness of people on local, national, and global levels. Global citizenship is a powerful concept that forces us to move beyond internationalization by acknowledging the links between the local and the global and by making explicit our underlying philosophical and theoretical perspectives (Hanson , 2009)

After that becomes apparent as nurse educators, we must continue the discourse around how to best situate these concepts within curricula, as well as examine the consequences for professional practice. The outcome of explicating what these dimensions might look like will promote a greater understanding of the contributions of nursing students and faculty as global citizens in the pursuit of health equity and social justice; provide guidance for nurse educators; contribute to the development of research outlines and inform theory development (Hanson L. 2009).

Theoretical Perspectives:

Although many authors have argued that there is a need to integrate social justice, global citizenship or social responsibility values into nursing practice and education, there has been limited exploration of the theoretical perspectives to guide this process. We believe that when engaging in global citizenship practice, education and collaboration, nurses, like any other health professional, must critically examine the philosophical underpinnings of their work. We highlight several theoretical perspectives and positions for consideration in the planning and integration of global citizenship into nursing practice.

A critical starting point in this process is the challenge of taken-for-granted assumptions and the clear articulation of the ethical and moral principles that will support individual and organizational activities. For instance, everywhere in the world, with marked differences between regions and countries, some social groups are excluded from achieving a healthy existence and their full life expectancy because of economic, social, and political factors, nurses could be contributing to injustice rather than advancing a social justice perspective. We believe that by adopting a clear standpoint promoting social justice, nurses can critically analyze the context from where they come and in which they will collaborate and practice (Hanson, 2009).

Global citizenship practice and collaboration should be based on a multiplicity of world views to guide the process and in employing a relational and flexible approach to identify and value complexity, the challenge resides in identifying unjust situations in the myriad of relations inherent to international projects and in operationalizing justice in a field that is permeated by many circumstances beyond the control of participants (eg, the structure of the health care system, university expectations, nurses' income, career paths, and personal circumstances, among others). A consideration of complexity, relational practices, and nursing ethics provides nurses the means to understand such experiences and to envision how one might integrate reflexivity, intentionality, and openness into nursing practice within a multifaceted global context (Keeping, Shapiro, 2010).

The identification of a theoretical framework from the social sciences will enhance the likelihood that critique it is integrated throughout the process of global citizenship nursing practice, education and collaboration. This framework opens space to examine the differences of class, gender, race, nationality, and other social relations as they impact individuals' and communities' health status. As a theoretical framework reveals how local and global issues are interviewed and how the lives of the majority of the world's population, including most nurses, are affected by economic, racial, and gendered forms of exclusion. This theoretical approach is considered appropriate for nursing given the profession's large female constituency and care giving (Boutain, 2008).

In Cairo universities and internationally many of the nurse scholars have critically examined the philosophical underpinnings of the concept of social justice and used a postcolonial feminist lens to bridge the gap between the speechifying of social justice and the reality of nursing practice. The use of such a critical lens is advocated to ensure that taken-for-granted assumptions, such as the superiority of professionals from high-income countries, are revealed in the relationships established. The certainty in the superiority of high-income countries can obscure the fact that the richer profession has been built through exploitive relations with other ethnic groups, locally and in other continents (eg, colony) (Boutain, 2008).

Therefore, race, ethnicity, and language are interrelated with many issues of privilege that are not perceived by most until they are enacted in practice; an example of such relations is the unrealistic expectation for partners to speak and write in their second or third language, while high-income partners function in their first language. In addition, many of our partners are part of the privileged in their home country. Often we do not critically examine the meaning for those involved and for those not contemplated in our collaborative activities and exchanges. We agree with many authors who emphasize that global citizenship aims to expand inclusion and power and provides the ethical and normative framework to make this a legitimate and far-reaching project whereby citizenship is a product of diversity rather than an institutional tool serving particular groups (University of Alberta, 2008),

The privileging of high-income countries as a reference point for work with the majority of the world is problematic. Most solutions created in high-income countries are not sustainable anywhere else in the world, and if the world had a more equitable wealth distribution, they may not be feasible in any country. A theoretical framework such as postcolonial feminism can signal potential pitfalls such as the use of Western solutions as potential models for LMICs. Such frameworks suggest the need for multiple forms of expertise, including local and national nursing associations and nurse academics in LMICs. Finally, we argue that there is a need for the constant exercise of reflexivity and use of mechanisms for formal and informal evaluation throughout the life of partnerships (Boutain, 2008).

Nursing Education Innovation Role in Nursing Citizenship

Nursing curriculum includes formal, structured and informal learning experiences provided to the different students. Planned components include theoretical courses and content, clinical experiences, and carefully constructed assignments. The exposure to scholars is formally rare, library holdings, faculty and peers, and voluntary activities. Thus, creating global Citizenship is multifaceted within the institution and the nursing education environment. It ranges from course offerings to diversity in the student population and the available clinical placements. One important factor contribute to theoretical understanding of global citizenship is the growing body of nursing literature on cultural competence and cultural safety. For example, Anderson and colleagues suggest that the concept of cultural safety encourages us to think critically about ourselves and to reflect on our historical, sociocultural, and economic locations (Faculty of Nursing, University of Alberta, 2006).

Situation that raises awareness of issues related to aboriginal health, immigrant health, international health, and other socially mediated inequality contributes to global citizenship. As Hanson states, the development of global citizenship must do more than create international placement opportunities or use global examples. For example, Mayo found that students who spend a significant period of time with a population at risk in a community setting develop a better understanding and appreciation for the concept of social responsibility, so we me merging he clinical practicum in to the theoretical part Thus, the main issue in this discussion of trends, challenges, and possibilities will focus on a growing clinical and curriculum response to global citizenship, the international clinical practicum, as an example of a strategy that is more complex than is often acknowledged (Faculty of Nursing, University of Alberta.2006).

International Clinical Practical

Since the mid-1980s, international practical have been advocated as one strategy to integrate a global dimension into nursing curricula. International experiences are beneficial for students to increase their understanding of the influence of culture on health, populations, to practice with diverse and to foster a global perspective on health. The development of knowledge about social justice issues must be combined with the opportunity for students to identify and integrate social justice principles into their clinical practice. Several different models may be used to facilitate the development of global citizenship: the exchange model implies the ability of both parties to reciprocate in the experience and ensures mutual benefit for both partners. The experience model implies a one-way opportunity for students (typically from *Global Citizenship and Nursing Education* high-income countries) to participate in an international experience in an LMIC; and the partnership model implies the joint development of courses and programs with experiences for students and faculty in both settings. Despite the proposed benefits of international experiences, there has been limited research to document the outcomes of international practical and experiences (Faculty of Nursing, University of Alberta, 2006)

Furthermore, there are several risks and issues that have been identified in the literature that must be considered prior to implementing international activities.

Resource, Practice, and Ethical Issues

Typically, international experiences have been resource intensive for faculty and universities alike. The costs associated with participating in an international experience are commonly borne by the students, and in some cases schools of nursing have contributed to the costs of sending students on an international experience. International *clinical* practical are associated with additional challenges and complexities because of the need to ensure that adequate clinical supervision is provided to students. Licensure to practice nursing is an ethical challenge related to international practical and exchanges. For graduate students and faculty to participate in international exchanges there is a need to ensure patient safety and maintenance of nursing standards. Typically, this is maintained through licensure. Differences across countries in relation to licensure requirements for such exchanges often reveal inequities when one partner allows clinical access and the other is constrained because of needs to comply with licensure rules that restrict access to certain clinically related activities (Faculty of Nursing, University of Alberta, 2006).

It is important to consider if the licensure requirements are the same for all participants in the exchange; if not, we must ask ourselves if the exchange perpetuates unequal power relations.

The transfer of nursing theories and models from "high income to many international experiences creates ethical and moral challenges that must be considered. This transfer may occur directly through the exportation of Western curricula and educators to LMICs, or indirectly through student and faculty exchanges. A few authors have argued that international practical have actually contributed to neocolonialism. Another pressing ethical challenge results when students and faculty from LMICs have decreased opportunities to engage in international experiences and exchanges due to the resources required to participate. The questions that become apparent in relation to international experiences are who participates, and more important (Canadian Nursing Association, 2009).

One of the most challenging dilemmas faced by faculty in the consideration of international activities is balancing responsibility for the health challenges in the local, national, and global arenas. Faculty who may not be fully committed to the need for the development of, and contribution to, global citizenship may be skeptical to commit resources to activities that do not contribute directly to local health challenges. Furthermore, the current shortage of nurse educators in both high- and low-income countries results in international activities being more difficult to implement. Despite this context, there are excellent examples in the literature of efforts to foster global citizenship by highlighting the link between local and global health challenges in coursework and programs. In an era of health challenges at all levels, and shrinking resources in many settings, what is our responsibility as educators to impart the sense of "being ethical in a global community" to our nursing students? (Hirschfeld, 2008).

Program and Academic Issues

Faculty who are involved with international exchanges can attest to the challenges that are faced when curricula lack the flexibility required to adapt to the needs of students from other settings. Differences in academic years, program requirements, and teaching approaches combine to make international exchanges, particularly at the undergraduate level, challenging to plan, implement, and evaluate. The quality and feasibility of international exchanges is also influenced by the lack of global educational standards. Several authors have highlighted the need to develop common educational standards to maintain global education and practice standards (Chen, and Berlinguer, 2001)

In relation to evaluation, there is a paucity of research on the best models for international exchanges, particularly clinical exchanges. Furthermore, in nursing there have been limited programmatic, research-based models on the development of global social responsibility. In the past several decades, the availability of international experiences for nursing students has been dependent on the values and attitudes, individual interests, and workloads of individual faculty members. With an increasing emphasis on the development of programs of research, there is a growing awareness of the need to integrate teaching and research activities within a program of research. Despite this awareness, there is limited funding for global nursing research. In addition, international work is often not recognized and valued within traditional university evaluation systems of faculty members (Chen, Berlinguer, 2001).

The planning and implementation of international practical and exchanges take additional time on the part of both faculty and students. These factors in combination have resulted in varying, and often limited, opportunities for students to participate in international experiences. While an important activity for universities, internationalization generally and international experiences specifically are insufficient as a mechanism for the development of global citizenship among students, faculty, and staff. International experiences are frequently the domain of relatively few students and faculty, are always resource intensive, and commonly fail to link the global with the local. We argue that a comprehensive approach is required to ensure that the development of global citizenship becomes a fundamental mandate of university programs. This approach must include the following key elements: Global citizenship content must be integrated into existing courses and through a specialized course to ensure in-depth expertise; faculty require professional development opportunities to ensure that they have the necessary knowledge and comfort to teach global citizenship; and global citizenship must be endorsed by leadership at both the faculty level and the university level (Chen, and Berlinguer, 2001).

Global Citizenship and Nursing Education

Global Citizenship and Nursing Education to health for all: that is they are expected to become global citizens. Becoming global citizens however does not imply that we privilege the global over the local, but that we recognize the interconnectedness between the local and the global. For example, over the past several decades we have become increasingly aware of the environmental, health, and migration issues that cross borders and impact all of us. International experiences and partnerships are powerful mechanisms to engage the next generation of nurses with the concept of global citizenship. The resources required, and the issues encountered to implement these experiences however, result in many of our students being unable to access these experiences and many of our faculty being unwilling to lead them. It is incumbent on nurse educators therefore to develop additional strategies to ensure that *all* students are exposed to, and have an awareness of, the concept of global citizenship and their role as professionals in a global world. In order to move toward this end, nurse educators and researchers must participate in the theorizing required on global citizenship in order to inform nursing curricula (Torres, and Rhoads, 2006).

Organizational citizenship behavior (OCB) comprises individual behaviors that are generally unnoticed but which collectively shape organizational orientation organizational citizenship behavior refers to the willingness of employees to go beyond the formal specifications of work roles, also known as extra-role behaviors. As noted by Baron & Greenberg (2008), a large proportion of organizational citizenship arises from informal behavior, with positive actions including voluntary engagement by employees beyond what is generally expected of them to contribute to the well-being of their organization. The individual who shows OCB performs more than required and expected, going beyond compulsory tasks identified formally by the organization. Organizational citizenship behavior is a profitable and efficient investment with positive outcomes for individuals, organizations and society as a whole. Literature regarding organizational citizenship behavior generally conceptualizes it in terms of five subscales (Baron & Greenberg, 2008).

Altruism involves voluntarily helping others with work-related problems, such as helping a co-worker with a heavy workload. Courtesy refers to gestures that help others prevent a problem, such as providing advanced notifications of meetings or of one's inability to attend them. Conscientiousness means exceeding the required levels of attendance, punctuality or conserving resources by not taking extra breaks and obeying company rules when no one is watching. Sportsmanship involves sacrificing one's personal interest and maintaining a positive attitude, even when inconvenienced by others or when one's ideas are rejected (Allen, Ogilvie, 2004).

Civic virtue involves the constructive participation in the political process of the organization, such as making suggestions for improvement in meetings Nurses perform better and exert high level of effort when they perceive that they are supported by organizations that care about their well-being and value their contributions, which encourages them to engage in organizational citizenship behavior. Organizational justice emerged as an important concept in the prediction of OCB, and some studies found a positive relationship between the two variables. To achieve OCB, justice must be generalized in the work environment. Perceptions of organizational justice constitute an important heuristic factor in organizational decision-making, as research relates it to job satisfaction, turnover, leadership, organizational citizenship, organizational commitment, trust, customer satisfaction, job performance, role breadth, alienation, and leader member exchange. In the same context, Ponnu & Chuah (2010) referred to organizational justice as achieving fairness within organizational settings, originating from work in social psychology aimed at understanding fairness (United Nations Educational, 2010).

Factors in Nurses' Organizational Citizenship Behavior

Issues in social interactions; Also, it is essentially the perception that subordinates of an organization are treated fairly. Such fair treatment can be manifest in numerous forms, ranging from the perceptions of the fairness of policies and procedures on an abstract level to the material distribution of rewards and punishments, along with a general ambience and working culture of being treated with courtesy and respect. Organizational justice remains a topic of deep interest for organizational researchers, as it represents an umbrella term used to refer to individuals perceptions about the fairness of decisions and decision-making processes within organizations and the influences of those perceptions on behavior. It is conceptualized as being comprised of at least two dimensions, the distributive and procedural while most recent research adds the interactional justice dimension. Distributive justice affects attitudes about specific *events* (e.g., satisfaction with pay, satisfaction with one's performance appraisal), whereas procedural and interactional justice affect attitudes about the *system* (e.g., organizational commitment, trust in authorities). Interactional justice can be further broken down into informational and interpersonal justice (Pike, 2008).

Organizational justice promotes an atmosphere of an atmosphere of trust, self-efficacy and confidence, especially in institutions with multidisciplinary staffs. Self-efficacy is important when engaging in proactive behaviors, as these behaviors entail certain psychological risks for individuals. Individuals who are confident in their capabilities are more prone to consider that their actions will be successful, and therefore assume the risk of being proactive. Self-efficacy beliefs link with high levels of taking charge and self-initiative, both of which constructs are similar to change-oriented citizenship behaviors. Such beliefs are related to role breadth self-efficacy (RBSE) which refers to employees" perceived capability of carrying out a broader and more proactive set of work tasks that extends beyond prescribed technical requirements. Extant research reports that RBSE is a strong predictor of behaviors such as suggestion making proactive behavior and proactive problem solving. Additionally, RBSE is an important predictor of employees" innovation and proactive performance. Indicated that self-efficacy is an important predictor of two types of proactive behavior: personal initiative and taking charge. Self-efficacy is clearly an important explanatory variable to consider when engaging in change-oriented citizenship behaviors (Mayo, 1996).

It were well to recognize this fact: that one of the things which has retarded many works for public good has been the existence of too many societies with similar aims, working side by side, dissipating great effort and much money in their rivalry and in their many machines. The one huge trouble at the bottom of all this misery we fight is the inconsistent behavior of man to man, our inordinate desire for money and our uncontrolled human passions. Together we must make radical changes. We need you and your support and influence. Our trained minds and bodies can offer you the best kinds of practical experiences and willingness to do what is needed (Abo Tiah, 2012).

The nurses who are responsible for the kind of material turned out from our training schools for nurses are those who belong to The National League of Nursing Education. They are the principals of these schools and the superintendents of hospitals; their efforts are concentrated on the best care of the patients in their hospitals and the proper education and development of their students. They try to make their standards of work, of teaching, and of living, as nearly uniform and as practical as possible. In the association with them are nurses who are doing educational work for public health, teaching the principles of rational living in any one of the many ways. A plea should be entered for these women, that they be not handicapped by political influences or by controversies among any of the organizations connected with the institutions in which they work (Abu Elanain, 2010).

The National Organization for Public Health Nursing is a body composed of many members. Modern medical science shows that it is poor business to cure people, only to have them reappear soon with the same malady, therefore, hospitals employ social service nurses to follow up their patients, observing their home environment and seeing not only that their families are taken care of by some outside agencies, but providing for the ultimate return to the wage-earner of business. The nurses investigate causes of illness, so that a patient's recovery may be permanent and that the malady may not recur among the family or neighbors. An effort is made to prevent any patient's being turned out of the hospital a wreck, to be cast about the city, an undesirable citizen (Ahmed, Fadel, Ghalla, Abo El Magd 2014).

Visiting nurses, as many of you know, receive calls from a central office and go out to visit patients in their homes. Doctors are called, if not already in attendance, and, if possible, financial aid is secured when that is needed. Sometimes patients may be removed to hospitals or other provision may be made for their care. When the patient remains at home the nurse calls daily (oftener, perhaps, in critical illness) and immediate care is given. Not only this, but instructions are always left with the patient for the protection of the family, and with the family for the care of the patient and of themselves (Ahmed, Rasheed, Jehanzeb, 2012).

Many a lesson in hygiene and sanitation is given while a bath is being prepared, and many a demonstration in home cookery as well as infant foods while the patient's diet is being directed. There are even visiting dietitians who go about teaching people with small means and little knowledge of such things, how to buy food and how to prepare it in order to feed their families properly. In many cities, milk stations are established where good milk is prepared by formula for sick or bottle-fed babies. Often there is a demonstration room where mothers come to learn how the food is made fit for their particular need and to learn the principles of hygienic and sanitary baby-raising. Who has not heard of the Little Mother Clubs, where little girls and older ones learn these same things, with the practical work of actually doing scientifically all that needs to be done for babies? Through such methods of education, infant mortality is already greatly reduced, though the work is only begun (Alabi, 2012).

The tuberculosis nurse follows much the same plan in her work for tubercular patients as do the hospital social service nurse and the visiting nurse in theirs, except that she sees only those with tuberculosis or those exposed to its infection and gives specific treatment and supervision. In tuberculosis hospitals and colonies, education and nursing go hand in hand. If patients do not infect others they may reinfec themselves, thus undoing all that has been done for them. Factory nursing and industrial hygiene are invested with much interest and opportunity for nurses. Factory owners and corporations are divided in their opinions of the value of these subjects. It will usually be found that those who break laws and are indifferent to the welfare of their employees are not long-headed enough to recognize the value of such innovations. However, where nurses are engaged they use the follow-up method; they look after those who become suddenly ill or who are injured. They try to prevent contamination of all the workers and to prevent infectious and contagious diseases (Al-Adel, 2001).

They anticipate trouble by looking after the anemic girl and the one who has persistent headaches. They do surgical dressings in the factory hospital office as they are needed. The man who goes home sick is visited there and the social service idea is fully carried out. In the factory, new co6peration is secured between employer and employee. While the workers are made more comfortable and are taught how to live decently and rationally, they in turn give better profit in labor. It has not infrequently been found that many improvements in machinery and factory equipment can be made to meet the health requirements as a result of the nurse's work in the factory. A kindred phase of nursing is that of looking after the employees and shoppers who become ill in large department stores. There is usually a very good emergency hospital with nurses on duty and perhaps a physician in charge. The prevailing opinion is in favor of such an arrangement as being of great value to the women and girls of these stores, for there is very likely to be greater care taken generally of their comfort and welfare and more thought given to justice and honest dealing in a place so regulated than elsewhere (Altınbaş, 2008).

Large telephone exchanges sometimes employ nurses who are responsible for the health standard of their employees. The girls are given rigid physical examinations of heart, lung and nerve resistance, as well as for minor though not less important details of health. Lectures pertaining to hygiene are given; the girls are made comfortable while on duty and a degree of attention is paid to outside influences brought to bear upon them. In some cities, nurses have allied themselves with bureaus for the protection of immigrants and have found a large field open for aid to foreigners. This work consists in finding homes and employment and the right social and religious environment for the girls and women. They need to be taught how to live in a new country; often former methods of living must be entirely given up, and so schools are formed and instruction is given in home nursing, housekeeping, cooking, hygiene and sanitation (Amikhi, 2009).

Citizenship and trustful relationships

Trust is crucial in all relationships formed within an organization, particularly in relationships between the staff and their managers. Trust, a crucial component of professional life has favorable consequences for both the staff and the organization. It is acknowledged as a factor that ensures that employees move toward a common goal and collaborate in pursuit of that goal. Managers, therefore, pay particular attention to developing trust among employees and working with employees who trust in each other. The concept of trust has been defined as: A feeling of confidence and commitment without the perceptions of fear, hesitation and doubt, where the person believes that he/she will receive support and collaboration in resolving problems in times of need without any underlying, ulterior motives and/or negative thoughts on the part of others. Organizational trust, which forms the basis of intra organizational relationships, has several definitions in the literature (Baron & Greenberg, 2008).

Organizational trust

According to Cummings and Bromiley (1995), organizational trust is The belief of an individual or a group as a whole that individuals or the organization will make every effort, whether explicit or implied, in good faith to act in accordance with commitments; that honesty in relationships will be ensured as a consequence of commitments; and that involved people will not seek to take advantage of others even if they have such opportunities. The term has been defined by Demircan and Ceylan (2003) as "the way an employee perceives the support offered by the organization, and his/her confidence in leaders or associates that they are honest and true to their word" (p. 142). Another definition provided by Yucel (2006) is "expectations of individuals, groups or organizations from individuals, groups or organizations with which they are in mutual interaction that they will make ethical decisions and will develop behaviors that are based on ethical principles" (Baron & Greenberg, 2008).

In research studies conducted on organizational trust, relationships between organizational trust and parameters such as performance, organizational commitment, job satisfaction and intention to quit employment, teamwork, and organizational citizenship behaviors (OCBs) have been investigated, and positive results have been obtained for organizations. These studies have demonstrated that organizational trust had a positive impact on employee performance (Demircan & Ceylan, 2003; Halis, Gokgoz, & Yasar, 2007; Laschinger, Finegan, & Shamian, 2001; Toprak, 2006; Yılmaz, 2006), confirmed that confidence among employees and toward managers increased job satisfaction (Laschinger et al.; Velez, 2006; Yucel, 2006), decreased absenteeism and intentions of quitting the job (Chen, Hui, & Sego, 1998; Yahyagil & Deniz, 2004), resulted in improved teamwork and team success (Erdem& Isbası, 2000; Toprak; Yılmaz; Yucel), increased commitment to the job and organization (Velez; Yahyagil & Deniz, 2004; Yucel), heightened motivation (Halis et al.; Toprak), heightened organizational commitment (Arı, 2003; Eser, 2007; Halis et al.; Ozer, 2007), and increased OCBs (Ylmaz).

"Good soldier syndrome," an OCB and a prosaically behavior seen in organizations, involves acts that are not directly and definitively described in formal job description, do not require penalty if not fulfilled, and are unplanned and not based on any directives or any perception of necessity but rather on volunteerism (Bolon, 1997; Cetin, 2004; Isbası, 2000; Podsakoff, MacKenzie, Moorman, & Fetter, 1990). OCBs, as a whole, have also been defined as voluntary actions that may include assistance, sharing, and contribution and that contribute to the effectiveness of the organization and are aimed at maintaining and protecting the peace among individuals and groups, as well as within the organization during the course of the fulfillment of organizational goals (Cekmecelioglu, 2007; Kose, Kartal, & Kayalı, 2003; Ozdevecioglu, 2003; Sabuncuoglu& Tuz, 2005). Five OCBs are described in the literature: altruism, conscientiousness, courtesy, civic virtue, and sportsmanship.

Altruism involves assistance provided by the members of an organization to others who have high workloads or who are experiencing problems with their work or to new members who have recently joined the organization. It also involves taking over responsibilities of others who have been absent due to illness. Conscientiousness involves behaviors such as consistent efforts to arrive at work on time even under adverse weather conditions.

Other examples include careful use of tea or coffee and lunch breaks, regular participation in the organization's meetings, and working unpaid overtime. Courtesy involves actions of an employee who asks opinions of others who might be affected by the decisions he or she makes. Courtesy also would include behaviors that safeguards the rights of associates, are constructive under any conditions regardless of the problems at hand, warns other members of the organization against potentially threatening actions, and tries to prevent problems before they occur or minimizes the possible effects of a problem (Cetin, 2004)

Civic virtue involves regularly participating in organizational meetings and discussions. Civic virtue also involves closely observing and trying to keep pace with changes in the organization, and taking active roles in relationship Between Organizational Trust—OCB assisting other employees in adapting to these changes, as well as suggesting solutions to problems and participating in the decision-making process. Sportsmanship involves avoiding negative behaviors that may result in stress as well as avoiding the exaggeration of problems. Sportsmanship includes avoiding focusing on what is wrong with work and avoiding disrespectful behaviors toward coworkers (Cetin, 2004; Isbası, 2000; Konovsky & Organ, 1996; Podsakoff et al., 1990; Sabuncuoglu & Tuz, 2005; Schnake & Dumler, 2003). Individuals working in an organization where a high level of confidence is present perceive themselves as a valued and important part of the organization, come to work with more enthusiasm, and are happier with their jobs (Arı, 2003; Islamoglu, Birsel, & B ̈or ̈u, 2007) As a consequence of this positive atmosphere, these individuals develop better relationships with their superiors, and they are more concerned with the success and future of the organization (Halis et al., 2007; Islamoglu et al.; Laschinger et al., 2001; Toprak, 2006; Yılmaz, 2006).

In addition, employees who have confidence in their managers, associates, and organizations are more apt to demonstrate behaviors of organizational citizenship, which include altruism, conscientiousness, courtesy, civic virtue, and sportsmanship (Cetin 2004; Demircan & Ceylan, 2003; Isbası, 2000, 2001; Laschinger et al.; Toprak). In studies conducted to determine the relationship between organizational trust and OCB, positive relations were found to exist between these two variables. In a study conducted by Dolan et al(2005) on 450 workers in Israel, a significant and positive relationship was found between organizational trust and OCBs. Podsakoff et al(1990) similarly reported that trust in the leader was influential on the staff's OCBs.

Other studies conducted on teachers showed that correlations existed between OCBs the teachers demonstrated and their trust in their institutions; that the teachers demonstrated voluntary actions more frequently as their trust in their managers, colleagues, students, and parents of students increased (Samancı, 2007); and that organizational trust was very influential in teachers' demonstration of OCBs (Polat, 2008). Samancı reported that teachers displayed more discretion and did not exaggerate the problems they faced and that their trust in the institution, being influential on their participation in extra activities, self-improvement, disregarding negative occurrences, and displaying helpful behaviors, also made them work more attentively. Teachers who trusted their institutions also put more effort into increasing the success of the school than those who did not. Another study conducted on 1,478 teachers concluded that teachers who willingly performed their duties had more trust in their schools compared with teachers who did their work unwillingly (Ylmaz, 2006).

In their study of OCBs in employees from different disciplines working in hospitals, Koberg et al(2005) determined that those with trust in coworkers shared their problems as well as their responsibilities with others more readily. In a postgraduate thesis study by Isbası (2000) examining the role of employees' trust in their managers and their perceptions of organizational trust in development of OCBs, trust in the manager was reported to be a mediating factor in developing OCBs. OCBs and organizational trust protect organizations from destructive and undesired behaviors and increase the tendency to collaborate and to share information among the staff, and to develop a sense of responsibility in the staff (Cetin, 2004). As a result of these positive attitudes and beliefs, motivation and satisfaction are improved. Thus, they have gained further significance in establishing more transparent communication and cooperation among members of the healthcare team and between the management and the staff, and have improved the quality of patient care and satisfaction (Ozdevecioglu, 2003; Velez, 2006). Therefore, when recruiting new staff, employers pay particular attention to select candidates who have high levels of organizational trust and who have the potential to fulfill the requirements of OCB in their institutions (Cetin, 2004).

In the healthcare industry, where a multidisciplinary approach and good communication, collaboration, and teamwork are essential, nurses are the healthcare providers who have the most familiar and most frequent interactions with patients. Poor organizational trust leads nurses to quit their jobs, which results in increased workload for other nurses in the institution. Increased workload may have adverse consequences, including interruptions in work, decreased motivation and performance, deterioration in the quality of patient care, absenteeism, and quitting (Laschinger et al., 2001; Velez, 2006).

Organizational trust is a key factor in reducing the rate of resignation of nurses, who are the healthcare professionals with the highest intention to quit. It is essential that nurses have confidence primarily in their managers, institutions, and coworkers in order to provide healthcare services in a more efficient way, ensure satisfaction in both patients and nurses, improve nurses' motivation and performance, establish their commitment to their organizations, and decrease turnover rates. Hospital management, on the other hand, should implement procedures to obtain predictive data by determining the level of organizational trust of the staff. Management also should take appropriate measures to eliminate the factors that lead to mistrust and measures to decrease the turnover rate and nursing costs (Altuntas, 2008).

Although job satisfaction, exhaustion, motivation, organizational commitment, and performance have frequently been described in previous studies in the nursing literature, the fact that the literature contained few studies on the concepts of organizational trust and organizational citizenship and the fact that no studies had been conducted regarding the relationship between organizational trust and OCBs in nurses, constituted the rationale for the present study. This study is important as it helps to emphasize the significance of this subject matter from the viewpoint of the management of the nursing services, and it helps to attract the attention of nursing services managers to the subject. Moreover, it is important that nursing services managers know nurses' levels of organizational trust and organizational citizenship as this knowledge helps them manage effectively by allowing for estimations in subjects such as organizational commitment, job satisfaction, performance, and intention to quit the job. Furthermore, this study contributes to the nursing literature and to the profession of nursing as it defines the relationship between organizational trust and OCBs in nurses (Stanley, 2017).

Improving health worldwide is a daunting task, and the obstacles are many. Daly points out that nurses are poised to meet the challenges of "providing care in under-resourced, chaotic healthcare environments; managing the increasing demands for nursing care in a context of work force shortages, dealing with increasing complexity in medical therapy, adapting to advances in technology, dealing with quality and safety issues, and the challenge of providing safe, quality, timely care." (Faculty of Nursing University of Toronto, 2010).

The key to future success, says Rafferty, will be in collaboration across disciplines and across borders: "We need to search for solutions to create sustainable welfare systems and seek out the evidence from best-in-class practice—not only from so-called developed but developing countries. Creativity and ingenuity are our currency for survival, and we need to encourage nurses to design new ways to deliver care with patients, to co-create new ways of tuning into patients' preferences, to be engineers of change. According to Daly, nurses will also need to "adjust to—and in some instances lead—advances in primary healthcare and new models of care even in acute care environments." It's a tall order. But will we have enough nurses to meet such problems head-on? Every global health leader we spoke with expressed concern about the global nursing shortage (Altuntaş, 2004).

A [silver] tsunami is developing," warns De Geest. Worldwide, our population of nurses—and patients—is aging. "We will be hit at a certain moment in the coming years, with workforce shortages being a major problem for all of us." "The global nurse shortage affects healthcare delivery in every corner of the world and will require interventions from all sectors of society, agrees Huijer. "There is also an increased demand for nurses with enhanced skills who can manage a more diverse, complex and acutely ill patient population than ever before." In response to the shortage, global nurses will look for ways to develop nursing capacity in every corner of the globe, says Michael Johnson. "If we want to teach sustainability and see true partnerships and relationships" when U.S. nurses conduct research and practice activities abroad, he notes, and efforts must "be related to capacity building in the country." (Altuntaş, 2008).

How can we create a sustainable global nurse workforce to tackle these challenges?

"Developed countries [like the U.S.] must be more attentive in exploring actions to stabilize and increase their domestic supply of nurses and moderate demand through strategic investments," says Huijer. "Even without the migration of so many qualified health professionals to work in developed countries, most less-developed countries do not have the healthcare workforce capacity to respond to the health problems of their citizens." The migration of nurses away from countries that need them most, says Huijer, "can threaten global health," which could make health, especially nursing, "a legitimate focus of international aid and democracy building. Quinn believes education, not migration, is the key. Developing nurse leaders, educators, and mentors, he says, will attract more students to the field. "In many countries nurses are relegated to a second tier and that needs to change. They need to be empowered to be the leaders of their own field and shape it for future generations to come (Altuntaş, 2008).

Caring for the world seems to be a unity task.

Woolley certainly believes so. "People want a connection to worldwide society, not just their own. I think we will see people make time for issues outside their current experience and expertise," she says. "The information environment we live in makes it possible to feel we are engaged with the globe anywhere, anytime, if we want to be." And nurses certainly do want to be connected, says Woolley. "Nurses are getting involved in the vision that global health is America's health. They are looking for ways to get involved." "We can make a difference for people in our own country and abroad," says Quinn. "My advice: Stay active, stay open to suggestions, and look to the world." (Altuntaş, and Baykal, 2010).

Let's start with the basics: What is global health?

According to the Consortium of Universities for Global Health, the term indicates an area of study, research, and practice that places a priority on improving health and achieving equity in health for all people worldwide. Anne Marie Rafferty's definition—"Global health is concerned with systems, the interdependence between them and how they impact on the experience of healthcare at local, national and international levels"—adds an interactive infrastructure to the definition and is further expanded on by Martha Hill: "It's [also]about long-term relationships that are mutually beneficial and addressing issues that span borders." Thomas Quinn notes that it "… incorporates multiple disciplines, interdisciplinary approaches to solving the health problems of the world. … It's not limited to one field. It belongs to all fields of expertise, directly or indirectly related to healthcare and health well-being for all people (Boerner, Dütschke, and Schwmmle, 2005).

The nurses and the global citizen health

"Nursing is integral to the definition of global health," says Quinn. "Nurses can play a more important role than just providing the care. They can help shape policy about how care should be given and develop best-case scenarios for improvement in life and building the health capacity of a country. But being a global nurse, he says, doesn't necessarily mean practicing nursing "beyond the borders of the U.S." It can also mean being a nurse in your own community. "It's still practicing health equity and that's the common denominator of the issues," he says. Whether working at home or abroad, having a global perspective—and experience—can offer nurses opportunities to grow and to serve, says Huda Abu-SaadHuijer: "International experience for nurses presents a powerful and rewarding option in addressing leadership development challenges, both global and domestic." (Boerner, Dütschke, and Schmmle, 2005).

Advantage of global citizen health for the nurses

Absolutely, it provides growth opportunities in terms of how we see ourselves, the world, and how we interact. Participating globally provides "stunning and exciting opportunities to learn, partner, innovate, collaborate, and build capacity, she says. Practicing abroad or working with international partners can enhance one's career, and you thrive with the intellectual and professional relationships and opportunities Sabina De Geest calls this "bringing scientific oxygen to the system." She believes that "by experiencing different systems, you will have a better appreciation of your own system. Experiencing these different systems, adds Mary Woolley, is an ongoing educational opportunity: "We can learn from experiences in a nation other than the U.S. and bring it [back home]. The learning can go the other direction as well. It needs to be constantly informing, so problem solving can take place (Cetin, 2004).

Role of the nurse in the international health collaboration

"The possibilities are endless for nurses in global health. They range from providing an educational role to doing research in a focused way," says Gail Cassell, alluding to a wide variety of opportunities for nurses to make a difference. The common thread among them, she says, is that "nurses bring a patient-centered focus to the global health team. Rafferty also spoke to the value that a nurse's perspective offers, Global nurses practice in a way that is consistent with the values, mindsets, and behaviors associated with global citizenship and play a role in leading change, promoting what I'd call cosmopolitan values in a sustainable way. Nursing has an "untapped potential," says John Daly. In his view, nurses are able to "develop and implement models of care which will contribute to the renewal and strategic development of sustainable, quality primary care. Well-educated nurses are well positioned to contribute to health system reforms and healthcare capacity development (Ciçek, 2010).

The role of nurses

How does the Faculty prepare students to be socially informed, well-educated global Citizens of the New world order? And what role can nurses Play to ensure that the world community delivers on its global health imperatives? As a leading research-intensive nursing faculty, we have been seriously considering these questions. One challenge that nursing faces is that it straddles the three domains of education, service delivery and professional regulation. To affect any change that has a lasting impact, all three elements need to be coordinated. If governments invest in education programs through universities and colleges but the education sector has no formal relationship with hospitals or other clinical sites, then the clinical aspect of the education will not improve. Similarly, if there is no role for the profession to set standards of programs, licensure or regulation, then it will be impossible to establish standards of practice or create quality expectations among educational institutions (Dolma, 2003).

In Canada, they have struggled with these tripartite complexities and benefit greatly from the fact that the professions are self-regulated but governed by law, that universities and colleges are strictly accountable to government for quality assurance, and that healthcare providers have rigorous standards of accreditation. It has been a long path to reach this point, and the standard of care that Canadians receive completely relies on these three components. Throughout the world, the development of a strong nursing profession faces myriad challenges. In some countries, war or political instability has undermined the regulatory framework, and "nurse" and "midwife" are not protected titles so anyone can use them. In other countries, nursing programs are new to the university sector; there is no history of collaboration between the clinical site and the university, making it difficult for the university to provide appropriate clinical experiences for students (Geçer, 2008).

In some areas, the clinical settings are so understaffed and under-resourced that students pose an impossible burden on the nurses practising there. One of the Faculty's goals is to meaningfully partner with colleagues in the profession, such as the Canadian Nurses Association, and with those in practice settings through the Academic Health Science Network (AHSN) and community health organizations. This goal ensures that the work we engage in internationally is not solely education focused, but pays attention to the professional and practice contexts. Sadly, in our experience these complex dimensions to nursing are not always evident to funders and policy-makers, largely because nursing is seldom well represented (if at all) around the tables where decisions are made on how to best address a country's health needs (Güler, 2009).

Hitting the target

Meeting the Millennium Development Goals requires not only the innovation agenda of breakthrough vaccines; it requires a sustainable healthcare system. A sound healthcare system requires infra- structure that supports quality education programs for all healthcare professionals, national standards and regulation, and an appropriate framework for clinical education. It is on these latter issues that the University of Toronto has much to offer. Not only do we have the benefits of an integrated approach to education and practice in our own programs, but our partnerships across the health sciences and with teaching hospitals and community service providers provide us with a broad array of resources for both education and practice based initiatives (Jahangir, Akbar, and Haq, 2004).

This is the basis of our approach to global engagement. We are interested in opportunities for partnership that support our colleagues around the world to build capacity in education, professional advancement and furthering practice. It is only through this tripartite mission that nurses can truly contribute to strengthening healthcare systems and that a sustainable workforce can be achieved. Critically, these partnerships also provide vital opportunities for faculty and students to Engage as global citizens, to learn from our colleagues, and to work together as Nurses on a global agenda to address healthcare and health workforce needs (Koberg, Wayne, Goodman, Boss. and Monsen, 2005).

When the Faculty engaged in the international domain during the mid-20th century, it did so from the perspective of the development model prominent at the time. The Faculty had the expertise, and nurses from around the world came to listen and learn. Western values shaped the Faculty's professional mission and vision, and the effective transport of those values was its measure of success. Today, the world has reshaped itself economically and politically, and the line between here and there has dissolved. Global citizenship now means doing as much learning as teaching, as well as taking responsibility for the state of the poor and marginalized in our own backyard. For our students, the key message is that the desire to listen and learn is more important than the illusion of having all the answers (Jahangir, Akbar, and Haq, 2004).

In some countries such as: Ethiopia faces some of the world's most serious political, economic and healthcare challenges. The burden of disease from potentially preventable diseases, such as hiv/aids, malaria and tuberculosis, is staggeringly high. Ethiopian nurses face insufficient resources, high nurse-to-patient ratios and inadequate policies to support safe practice. These unfavorable working conditions have led many Ethiopian nurses to move to wealthier countries to practice, increasing the African nation's nursing shortage. Since 1999, Addis Ababa University, in Ethiopia's capital city, has graduated more than 6,500 nurses and midwives. In 2003, it upgraded its nursing course from a diploma to a degree program. Then in 2005, it introduced a master's of science (nursing) program. This rapid growth has created a significant demand for nursing instructors with postgraduate degrees and scholarly skills. But financial and material resources are scarce, making it difficult for the university to recruit faculty to fulfill its education and research activities (Kise, Kartal, and Kayal, 2003).

Countries Example to be Stronger Nurses Together

Nurses have a way of coming together to help each other out. The Bloomberg Faculty of Nursing and the Canadian Nurses Association (cna) have begun to link a number of initiatives to provide an integrated approach to working with our Ethiopian colleagues in the areas of education, practice and professional leadership. Building on can's strong relationship with the Ethiopian Nurses Association(ena) and the University of Toronto's links with the Addis Abba University (which includes medicine, pharmacy and engineering), the Bloomberg Faculty has connected the professional advancement project with professional education initiatives and joined forces with the Centralized School of Nursing at Addis Ababa University. Together, the cnaena, Addis Abba University and created the Ethiopia-Canada Nursing Collaboration which has developed a multiple-intervention approach to strengthen nursing in Ethiopia and support the efforts of Ethiopian nurses to improve their country's healthcare system (Organ, Podsakoff, and MacKenzie, 2006).

In April, Amy Bender, PhD 0T9, an assistant professor at the Bloomberg Faculty, travelled to Ethiopia with Angela Cooper Brathwaite, PhD 0T4, a Bloomberg assistant professor and a manager of injury prevention in the public health division of the Durham Region Health Department. They spent one month facilitating research and clinical leadership seminars and working with Addis Ababa University faculty to provide second-year graduate students with one-on-one thesis support. Further trips are planned for the fall and in 2011. "The faculty at Addis Ababa already teaches a good curriculum," says Bender. "The collaboration is intended to support the faculty in building the research and leadership capacity of their master's students, many of whom have limited clinical experience and yet are responsible for teaching nursing in other parts of the country. "The Bloomberg Faculty's immediate focus is on graduate education, but it will also work on creating links between advanced practice clinicians in Toronto and the ena, as well as nursing leaders in clinical settings in Addis. The motivation to improve nursing in Ethiopia is strong, even though the country faces many challenges. Bloomberg Faculty of Nursing will continue to stand with and support Ethiopian nurses through this time of growth (Zee, 2009).

Taking the lead

In an effort to address global health needs and meet the World Health Organization (WHO) Millennium Development Goals, leaders around the globe are working together to strengthen healthcare systems and create human resource capacity. Nurses are key contributors to this healthcare system research, planning, design and delivery. As well, they're key facilitators for collaborative relationships within and between countries around the world. It should come as no surprise, then, that many of the leaders in global health care were nurses (Zee, 2009).

The past several years have taught me that nurses have the requirements necessary to address global health needs by facilitating international collaborative partnerships. To approach these challenges, you require an open mind and flexibility to understand, appreciate and learn from cultural differences. You need resilience, creativity and resourcefulness to create effective partnerships and exchange knowledge. You also need tenacity, intelligence, conviction and a passion for improving global health outcomes. Nurses have these qualities in spades. From the beginning of my nursing career, I have worked to build nursing and healthcare knowledge focused on the health needs of people and populations while acknowledging concurrent social, historical, political, cultural and economic complexities that frame healthcare systems. As Madeline Leininger, who founded the field of transcultural nursing in the 1950s, contends, nurses are ethically mandated to use their knowledge and skills to promote global health (Park, Yun, and Han, 2009).

Nurses need to create actionable strategies for equitable, socially just global health systems that reflect the health and cultural needs of populations. Such strategies include health research partnerships within and beyond Canada. Global nursing leaders must believe in and foster the research capacity of others through mentorship and mutual learning. In so doing, they can create successful, sustainable research programs in Canada and in their partner countries. These programs, in turn, can lead to better healthcare. For me, creating global research connections is crucial to addressing the global-health and health-workforce crises, which include an inequitable distribution of the health workforce and a global deficit of healthcare providers that's currently estimated at more than four million. Africa, for example, bears 24 per cent of the burden of disease worldwide yet has less than two per cent of the global health workforce (Sabuncuoğlu, and Tüz, 2005).

The continent is further challenged by a poorly distributed healthcare workforce, high rates of attrition and an increasing prevalence of disease. As director of the American Health Organization Collaborating Centre on Health Workforce Planning and Research, he is privileged to collaborate with key domestic and international stakeholders in health policy, research and practice. The support of Health Canada, who, paho, Dalhousie University and the Canadian Coalition for Global Health Research has allowed the team to develop partnerships with Brazil, Jamaica and Zambia. These partnerships aim at enabling capacity in human resources for health research, research use, strategic planning, evaluation and knowledge translation. U of T Professor Linda O'Brien-Pallas and I worked with a team to develop a methodology based on an established needs-based health human resources conceptual framework. The collaborating center works with its three partner countries to tailor the framework to various governments, socio-economic situations and population health priorities. The emergence of a coherent communications strategy for sharing good-news stories with stakeholders in Jamaica and Brazil as well as here in Canada is one vital area where much growth is taking place. For example, based on its successes, Jamaica is working with other Caribbean countries and the collaborating centre to develop a Centre for Excellence in Health Workforce Planning and Research (Schnake, and Dumler, 2003).

In recent years, some cooperative behaviors that are not included in the formal definition of job and aims have emerged among co-workers of organizations, and these behaviors have gradually gained much more importance. The organization members perform these behaviors during their fulfillment of organizational roles in order to improve peace environment. Organizational Citizenship Behavior (OCB) is one of the extra role behaviors, which is also known as good soldier syndrome.

Organization Citizenship Behavior (OCB)

OCB is not directly defined and clearly indicated in the definitions of formal roles. OCBs are positive social behaviors that increase the efficiency of organization as a whole. These are the behaviors displayed voluntarily by workers depending on their personal choice without a written rule, and they are not clearly indicated in the agreement of an organization and not required by job definition; therefore, omission of these behaviors does not bring any penalty Five organizational citizenship behaviors are defined in the literature which are known by different names.

Courtesy - Based Information (Courtesy)

Courtesy stands for the positive behaviors displayed by the members who should be in communication within the organization and are affected by each other's works and decisions. Courtesy based information includes future-related behaviors like informing others before starting an action, warning other people in the organization about dangerous activities, and taking precautions to prevent or alleviate the adverse effects of problems supporting the development of organization (Civic Virtue): Civic virtue is the active participation of workers in the political life of the organization. For instance, it includes some behaviors like regularly attending intra-organizational meetings and discussions, closely following the changes in the organization and offering solutions to problems, and joining   in the   decisions made within the organization tolerance (Sportsmanship): Tolerance is the avoidance of negative attitudes that could cause tension between members of the organization, unnecessary discussion, spending too much time speaking of the problems at work, and disrespectful behaviors towards work partners (Anthea, and Pauline, 1977)

Organization Citizenship Behavior in Nursing Services

Nursing is a profession that most closely knows and interacts with patients in the health sector, and therefore, it requires the cooperation of more than one health staff, good communication, and team work. In this respect, OCB increases the inclination towards helping and sharing information, and promoting the feeling of responsibility, motivation, and satisfaction within the organization. For these reasons, OCB is indispensable for establishing a more precise communication and cooperation among health staff, managers, and workers in order to increase work quality and patient satisfaction the nurses not displaying OCBs such as helping, informing, tolerating, praising the institution, conscience are more inclined to demonstrate negative behaviors of reduction in the service quality, non-sharing among team members, and conflict. This could cause job dissatisfaction and reduce organizational commitment among nurses. Decreases in job satisfaction and organizational commitment will eventually cause nurses to quit their job. Besides potentially negative conditions will damage the institutional image, and therefore, they should not be preferred by individuals and institutions (Boerner et al., 2005).

Nurses who are the most active members of the health staff should display high levels of organizational citizenship behavior in order to more efficiently provide health services. Certain factors like nurse and patient satisfaction, motivation, and performance of nurses promote organizational commitment. Nurses' job satisfaction and commitment to institution reduce the turnover rate (Bolon, 1997). In parallel with the rapid change and development in the health sector, future staff should meet the demands and expectations of patients and have high team spirit. They should be able to work together with others in coordination and perform the requirements of organizational citizenship behavior in their institutions (Altuntaş, 2008).

Organizational leaders and nurse managers should determine the organizational citizenship behavior levels of nurses and effective factors, and then take the necessary precautions against negative factors, and encourage positive factors. Thus, they should increase nurses' motivation, job satisfaction, organizational commitment and productivity levels. From this respect, this study will provide information especially to the managers of nursing services about OCB levels of nurses and help them to make prediction for organizational commitment, job satisfaction, turnover intention etc. and develop the competencies of nurses. At the same time, the present study will determine the factors effective on OCB and help nurse managers to take necessary precautions for effective factors. Furthermore, it may provide guiding principles to researchers who want to study on this topic.

Reconceptualization of the citizenship among mentally disabled:

Historians know little about the specific contexts within which these issues were mobilized. Academic commentaries on the mental health charity MIND's activities during this time have concentrated on its rights-based campaigning regarding mental illness. Yet, MIND was also prominent in campaigning to change government policy and professional practice for people then called mentally handicapped. There is an almost complete lack of reference to the organization in contemporary historical overviews of these changes (Maria and Michael, 2016).

Critics of MIND's rights based approach, (focusing on its campaigning regarding mental illness) have commonly maintained that it was individualist in nature, and fundamentally opposed to medicine and psychiatry. It is claimed that debate was framed in terms of the deprivation of liberty. This enshrined personal liberty at the cost of care and treatment. The characterization has, in turn, been used to support the view that MIND encouraged patients' discharge from hospitals without enough consideration of the consequences for people's lives in the community. This article contests these representations. It argues that a proper appreciation of MIND's rights-based campaigning requires its contextualization with the organization's much longer history, which stems back, ultimately, to 1913. This history is itself intimately related to notions of citizenship. Its examination places familiar touchstones in the history of people with learning disabilities in a new light.

Under the arresting anthropological concept of 'soul catchers', a recent special edition of Medical History has examined the relationship between technologies of the mind sciences and those immaterial aspects of what, in the modern world, has become known as subjectivity (Maria and Michael, 2016).

This article does not directly reflect on such relationships. Instead, it attempts the more modest task of tracing how a conjunction of the altered socio-political terrain of the post-war Welfare State, with changing approaches in the medical and allied sciences, gradually transformed conceptualizations of learning disabled people's subjectivity and citizenship. Other recent articles in Medical History have reinvigorated Roy Porter's mid1980s clarion call for a 'patients' history from below', examining in detail elements of the so called 'doctor-patient' relationship and its relation to the construction and negotiation of medical knowledge in the mental health services. They trace the intimate connections between the mental hygiene movement's policies towards people then termed 'mentally deficient' and notions of citizenship. The third section examines MIND's transformation into a rights-based campaigning organisation and the associated reconstruction of the relationship of citizenship and learning disabled people. Its leading members comprised doctors and other campaigners who had been involved in pressing for this legislation. People considered 'mentally deficient' were considered a 'social problem' requiring control as well as care. Under the 1913 Act, all county and county borough authorities in England and Wales were to ascertain the number of people deemed 'mentally defective' and arrange institutional provision or community supervision (Maria and Michael, 2016).

Doctors received a key role in diagnosis and certification procedures. Local voluntary organizations could also appoint themselves to identify and supervise people. The CAMW set itself up as the central training and coordinating body for these organizations. Historians have commonly attributed the construction of 'mental deficiency' as a major social problem to eugenic concerns about national fitness. But Mathew Thomson has situated it within the wider political context of 'adjusting to democracy', convincingly showing an intimate connection with re defining rights and citizenship. The franchise was greatly widened between 1867 and 1918, despite continuing limitations (notably the exclusion of women under the age of thirty). This increased anxieties among some about the requirements of responsible citizenship. The 'social problem' of 'mental deficiency' became a focus for these concerns. As Thomson puts it, 'The category of mental deficiency. The mental hygiene movement began in the United States in the first decade of the twentieth century. On this history, see, for example, Johannes Pols, '"Beyond the Clinical Frontiers". It was originally called the Central Association for the Care of Mental Defectives, changing its title to CAMW in 1922 (Brian, 1994), Provided a way to conceptualize a group within the population who were non-citizens, not on grounds of wealth or class but because of an innate deficiency and social inefficiency.

'The CAMW's association of mental deficiency with 'social inefficiency' confronted it with people on the so-called 'borderline' between apparent pathology and mental health. Through this, the CAMW combined with the NCMH, the CGC and Tavistock Clinic in the inter-war formation of a movement for mental hygiene. This movement continued to campaign on the 'problem of mental deficiency' but it focused particularly on what it considered psychological causes of 'social failure' in the wider population. This focus was strongly informed by psychodynamic thinking. An emphasis on 'emotional adjustment' in the interests of adequate citizenship and social efficiency developed – with this construed in terms of mental health. This entailed, in particular, sensitive attention to relationships of nurture and authority in childhood upbringing (Maria and Michael, 2016).

Mentally deficient children's emotional lives and behavior difficulties were, however, largely considered mere consequences of intellectual incapacity. The mental hygiene movement pressed for greater institutional provision for mental defectives as a means of social and mental hygiene. But there was growing recognition that relying solely on institutionalization was not economically viable. Instead, an integrated system of institutions and community supervision was promoted. This 'community care' was often called 'community control', its nature conditioned by the original segregationist designation of 'mental deficiency' as the antithesis of citizenship (Brian, 1994),

Indeed, the families of 'mental defectives' remained suspect in terms of heredity, as well as practical and moral training. The inter-war mental deficiency system represents an unprecedented coercive and interventionist strategy for 'public welfare'. It was premised on mentally deficient people's supposed threat to the community, and their inability to perform the role of self-sustaining, responsible citizens. The Post-war Mental Hygiene Movement Post war, the conceptualization of mental deficiency as the antithesis of citizenship began to unravel. However, many of the assumptions associated with it continued to inform both the mental hygiene movement and the mental deficiency system. the leading post-war voluntary organization working for mental hygiene. Its Council comprised mostly doctors and other professionals representing professional organisations (NAMH, 2013),

This amalgamation took place alongside the newly elected Labour Government's institution of a Welfare State that was intended as a final break with the Poor Law. Social protection was to be a right of citizenship with universal and freely accessed services. But, as Mathew Thomson has noted, 'the impact of 'democratization', 'universalism', and 'social citizenship' was mediated by status'. The NHS shifted power to central bureaucrats and the medical profession. The principal role of doctors and medicine was reinforced under the NHS Act. Institutions became hospitals. Their administration became separated from community care provision. Local authorities were given permissive powers to provide services in the community. But, as they had formerly operated the institutions, community provision was minimal, and with powers only permissive, little was achieved. Meanwhile, largely in response to the war, the mental hygiene movement's concern about the social 'threat' of mental deficiency receded. Instead, it increased its focus on emotional maladjustments in the wider population. Nevertheless, the movement did not fundamentally question the role of the mental deficiency system. People deemed mentally deficient continued to be considered incapable of the social responsibility and citizenship that mental hygienists associated with mental health. Throughout the 1940s and 1950s, the mental deficiency system remained coercive and custodial. Criticism came from elsewhere, however. The National Council for Civil Liberties' (NCCL) lengthy campaign against the workings of the Mental Deficiency Acts is well known (NAMH, 2013).

Although the NAMH was aware of the long waiting-lists, lack of beds, and general poverty of conditions and staffing, it publicly refuted the NCCL's claims. It maintained that the NCCL was prejudiced, and its focus on 'wrongful detention', outdated since the modern mental deficiency institution, was not based on 'permanent detention' but on training. However, reaction behind the scenes was more complicated. Conflicting views found common ground in the need to free up places in overcrowded institutions. The NAMH decided that greater hostel provision and more generous licensing might achieve this and began to look at possibilities for amending the legislation in general. But a number of other factors were also clearly prompting the NAMH to alter its position. One was the growing influence of psychologists. Although hardly any were directly employed in mental deficiency hospitals, from the early 1950s a number began experimental research within them. This played a significant role in psychologists' (NAMH, 1971).

Assertion of a professional role during the 1950s and 1960s. It was also made use of by the NCCL. Results demonstrated that the great majority of people in institutions were no danger to society. They also revealed that IQs were often higher than supposed, and that people categorized 'feebleminded', as well as those considered 'imbecile', had a higher capacity to learn and work than the results of the current system suggested. The NAMH, in fact, helped publicize this research, inviting speakers to its Annual Conferences and publishing articles in its journal Mental Health. Indeed, the psychologists involved in this research increasingly worked with the NAMH, promoting new forms of care and rehabilitation. The NAMH was therefore more amenable to reforming the mental deficiency system than its public dismissals of the NCCL's allegations implied. Another factor was that the mental hygiene movement had itself developed a critique of institutionalization. This was derived from the movement's psychotherapeutic focus on mental health in terms of emotional adjustment and maladjustment. Increasingly, enforced living in large groups within institutions, as well as rigid authority and hierarchy, was considered to be detrimental to human relationships, and therefore to the emotional adjustment of inmates. Instead, the movement began to encourage more open communication and greater freedom of choice within a more egalitarian structure (NAMH, 1971),

These principles informed the mental hygiene movement's promotion of 'social therapies' and 'therapeutic community' ideas intended to modernize mental hospitals. They also informed the movement's strong influence over the 1945 Curtis Report on the Care of Children Deprived of a Normal Home. This had set the foundations of post-war policy for the residential care of children. It considered many existing institutions authoritarian, and therefore insensitive to children's emotional needs, while life in large groups, in particular, created emotional and behavioral problems. But the psychotherapeutic ideas that underpinned these principles had a strong developmental component that explicitly discriminated against most people considered to be mentally deficient. Therapeutic community approaches were not thought to be appropriate for mentally deficient people, and the Curtis Committee considered mentally deficient children to be outside its remit. Nevertheless, in the context of the NCCL campaign, the prominence of such principles reinforced the view that the mental deficiency system was out of date and custodial (Marshall, 1950).

Citizenship and Learning Disabled People responsive to parental needs

The NAMH developed a relationship with the NAPBC during the 1940s and 1950s. The impact of all of these factors on the relationship between citizenship rights and a diagnosis of mental deficiency was ambiguous. While the NAPBC pressed for wider service provision and, increasingly, sponsored research experiments, it was effectively the views of parents rather than mentally defective people themselves that were becoming more prominent. Meanwhile, under a post-war settlement that relied upon 'full employment', psychologists' research on IQ, learning capacity and ability to perform industrial tasks implied the potential reinstatement of citizenship for many people detained in institutions. However, it also suggested that citizenship rights were dependent upon an ability to perform often repetitive and poorly remunerated work (Marshall, 1950),

Indeed, there remained a sense that citizenship was dependent on the capacity of individuals to be improved in order to work and assimilate with the wider population. Similarly, ideas on emotional wellbeing and development were largely limited to 'high-grade' patients and translated into a need for training in 'socialisation'. Nevertheless, other emphases were also present. Tim Stainton has pointed out that the idea of social rights under the Welfare State certainly contributed to the influence of the NCCL's campaign. Society AGM, that the concept of 'high-grade mental defect' should be abandoned. He argued that these people would be better accommodated under the general post-war welfare services legislation. Indeed, he went further, arguing that the 'Education Act should be extended to cover "imbecile" children thus imposing on the educational authorities the responsibility of providing educational training for all children who can benefit from it'. The 'concept of normality should be broadened', he said. Two years later, the NAMH published the talk in its journal Mental Health (NAMH, Annual Report 1965).

The 1957 Report of the Royal Commission on the Law Relating to Mental Illness and Mental Deficiency expressed much of this ambiguity regarding mental deficiency and its relationship with citizenship. Although its appointment in 1954 had been spurred by the NCCL's campaign, it was also considered an overdue attempt to realign the mental illness and deficiency legislation with the post-war Labour Government's introduction of comprehensive and freely accessed Welfare State services. The Commission's terms of reference were to make recommendations on the possibility of treating patients informally without certification. This was a principle around which the NCCL and the NAMH could, in fact, find common ground. While it was critical of the diminished legal safeguards for admission, the NCCL nevertheless welcomed the emphasis on voluntary admission with its reduction of the penal image of care and treatment. The NAMH also welcomed this emphasis. In fact, its precursor organizations, along with the wider mental hygiene movement, had advocated this before the war. Yet this had been on the grounds that the necessary institutionalization of mental defectives was hampered by certification procedures that pandered to concerns about the liberty of the subject. Nevertheless, the post-war Welfare State emphasis on universal, freely accessed health and welfare services provided a new context for the NAMH's promotion of voluntary admission. Similarly, it was within this context that both the NAMH and the NCCL were able to support the Commission's assumption that psychiatry should be assimilated with general medicine (NAMH, Annual Report 1965).

This meant, however, that mental deficiency (re-labelled mental subnormality by the Commission) was subsumed under the term 'mental disorder', and remained within the conceptual and practical setting of 'illness' and 'treatment'. Nevertheless, the Commission accepted that there were debilitating effects caused by institutionalisation. It maintained that this should be avoided by making community care the preferred option with hospitals taking patients only when necessary in the interests of treatment. Indeed, in accordance with the general thrust of post-war Welfare State services that aimed to maintain citizenship by their universality, non-stigmatising and non-segregationist nature, the Commission's proposals were formulated into a policy that envisioned a major shift towards community integration of large numbers of patients (Maria and Michael, 2016).

The ambiguities regarding the citizenship status of mentally subnormal people inherent within the Royal Commission's proposals were enshrined within the 1959 Mental Health Act.

Consequently, they continued to be expressed through the 1960s. The NAMH welcomed the Royal Commission's Report but was concerned that the proposed wide extension of community care services should have adequate funding, trained staff and an efficiently coordinated overall system. It was disappointed, in particular, that the ensuing 1959 Mental Health Act did not make the development of community care services mandatory on local authorities as the Commission had proposed. The NAMH pressed for investment in community care services and staff training throughout the 1960s (Maria and Michael, 2016).

Yet, in terms of the relationship between mental sub normality and citizenship, the NAMH's views were equivocal. For instance, psychologists became more prominent within the NAMH during the 1960s. The informational booklets they produced emphasised that the needs of mentally subnormal people were social, educational and occupational, rather than primarily medical. However, those on 'social training' also emphasised training people so that they could assimilate as inconspicuously as possible. Citizenship rights were, however, implicit in one important research experiment funded, not by the NAMH, but by the National Society for Mentally Handicapped Children (NSMHC, formerly NAPBC). The influence of the Brooklands experiment on child (and adult) care is commonly cited. Yet, commentators have not drawn attention to the equally important assertion of citizenship inherent in its research design. Headed by Jack Tizard, Brooklands was a home set up for 'imbecile' children then living in the

Fountain mental deficiency hospital. It explicitly applied the principles that had informed the 1946 Curtis Committee on the Care of Children without a Home. Close, affectionate care was pursued and continuity of relations between particular staff and children attempted (NAMH, Annual Report 1965).

Emphasis was placed on the children's existing emotional needs, rather than the then prevailing attention on education and training. This effectively detached the traditional mental hygienist linkage of the means of care and treatment from the aim of producing socially 'responsible' and economically productive citizens under the banner of 'mental health'. These 'imbecile' children were unable to attain this goal, but the inherent assumption of the research design was that this should not be allowed to deny their right to the same quality of care and support as other children. More fundamentally, Brooklands foregrounded similarities of emotional experience and response instead of difference and deficiency. This in itself was a powerful statement of shared humanity and citizenship. It bears pointing out that the staffing and material provisions made available for the experiment itself were inadequate, especially in its first year. However, conditions at the hospital from which the children had come were themselves poor, with harassed nurses working on wards of around sixty beds. Indeed, these conditions were commonplace at mental deficiency hospitals at the time. Potentially more disquieting is the lack of information about the fate of the children after the experiment. It appears unlikely that there was a continuation of the care that it was asserted they were entitled to (Jahangir, Akbar, and Haq, 2004).

Notwithstanding these reservations about the experiment in practice, Brooklands inherently asserted citizenship rights. And it did so, on the basis of mental hygienist psychotherapeutic theories that had originally excluded people considered mentally deficient. In the 1960s, Tizard and colleagues, Norma Raynes and Roy King, extended the Brooklands research, highlighting the poverty of 'institutionally oriented care' in contrast with 'inmate oriented' care based on the Curtis-style 'family' model.

The NAMH began to criticise the management of some hospitals. In 1967, it called for more research on the workings of the Mental Health Review Tribunals (MHRTs), and began working with NCCL to design a patient representation service. The Emergence of MIND and its Rights-based Critique The NAMH's admission was prescient. A series of hospital scandals broke out in 1967. Alongside the developing critiques of hospital-based services, these pushed the NAMH into an even more critical and assertive position. Sans Everything, a book with a foreword by the psychiatrist Russell Barton, a prominent member of the NAMH, alleged serious abuse and neglect on wards for elderly people in several hospitals (Jahangir, Akbar, and Haq, 2004).

A scandal emerged at Ely Hospital in Cardiff involving abuse, along with generally poor treatment and conditions. Further concerns about conditions and treatment in hospitals followed the next year. In 1969, Richard Crossman the Secretary of State for Health and Social Services, had the official report on Ely published in full, despite resistance from his own Ministry. That same year, the sociologist Pauline Morris published a large-scale survey of thirtyfive mental subnormality hospitals funded by the NSMHC from 1964. This described severe overcrowding in often dilapidated buildings, with very poor staff–patient ratios, poor training and inadequate staff communication (Güler, 2009).

This would ultimately transform its original mental hygienist equation of mental deficiency with the antithesis of citizenship into an approach that placed the citizenship of people with learning disabilities at the centre of community-based health and welfare services. In 1970, the NAMH announced its intention to begin a major national campaign. It aimed, in particular, to highlight the problems of institutional care and of insufficient provision within the community. Regarding mentally handicapped people (as they were now commonly being termed) the NAMH maintained that: Our main concern is with the individual and his mental health, whatever his innate intellectual capacity. That the ability of the mentally handicapped to enjoy life should not be impaired by a lack of human warmth, appropriate assessment and every opportunity for self fulfilment. This statement marks a milestone in the NAMH and the mental hygiene movement's policy. It is a tacit recognition that intellectual capacity cannot, in itself, be a measure of mental health. Simultaneously, it represents an acceptance that the mental hygienist equation of mental health with adequate citizenship and 'social efficiency' had become subverted (Güler, 2009).

By the early 1970s, activists were attempting to give legal aid a principal role in gaining access to justice for poor and marginalised people. Legal and welfare rights were asserted by the emergence of self-help and pressure groups such as Tenants' Associations and Claimants' Unions. The Child Poverty Action Group (founded in 1965 in response to the Labour Government's failure to increase family allowances) opened a legal department in 1969 and, in 1970, a Citizen's Rights' Office. During the 1970s, it turned to litigation as a means to test the interpretation of social welfare law. Under its Director Tony Smythe, the NCCL promoted legal aid and advice for marginalised and disadvantaged groups in the later 1960s. As noted, in 1967 the NAMH and the NCCL had attempted to combine their expertise in the creation of a MHRT representation scheme. It is within this complex that the emergence and mobilisation of the NAMH's rightsbased thinking needs to be situated. For example, in 1972 it collaborated with the Disablement Income Group (DIG) and the Spastics Society to press for patients to receive pocket money as of right under social security legislation instead of at the discretion of hospital authorities (Geçer, 2008).

But the NAMH struggled to appreciate the underlying issues regarding citizenship and associated questioning of the social organisation of health and welfare services. This conflict can be illuminated by examining the NAMH's position in contrast with the Campaign for the Mentally Handicapped (CMH). The CMH emerged from the Guardian welfare journalist Ann Shearer's experience of visiting mental handicap hospitals. Her response was to prepare a manifesto calling for their closure. This drew hostility from many psychiatrists. But, with the support of Anita Hunt, a researcher at the Spastics Society, and Sandra Franklin, an architect, the CMH was launched in early 1971(Geçer, 2008).

In the same year, the CMH sent a document of its aims (probably Shearer's 'manifesto') to the NAMH asking for its support. The NAMH felt unable to approve the central proposition that hospitals should be closed, but sympathised with other aspects of the document. In fact, it was willing to distribute, and reprint in its magazine, a pamphlet called 'A Right to Love?' by Shearer which criticised institutional living and segregation. But the gulf between the CMH and the NAMH can be appreciated in the contrasting views on what to do about services for mentally handicapped people published in the first issue of Apex, the journal of the newly founded Institute of Mental Subnormality. Mary Appleby, the NAMH's General Secretary, forcefully attacked the polarization of views between the 'hospital lobby and the lobby in favour of community care'. She argued that there should be 'a single focus of leadership' which would transcend. Artificial distinctions between hospital and community, between treatment and care'. But this was to be done by enabling hospitals to regain the leadership role that they had lost. This 'power unit' would be the 'centre of a series of concentric circles of complex supportive provision'. Here the psychiatrist would be able to call on the skills of other medical, psychological, educational and social workers. It would 'concentrate the sum of technological knowledge on the most difficult patients', guide medical treatment, devise training programs and educate parents and staff. Appleby argued that this should be a completely separate service for mental handicap, able to command financial support from government through a single channel (Ciçek, 2010).

Ann Shearer's combative article offered a very different vision. She maintained that people's attitudes to mental handicap mirrored their deeper attitudes to society. Confronting them would force us to ask whether this is to be a society which accepts people of different abilities and the economic dependence which goes with this, or one which shuts them away from its normal patterns in however kindly intentioned a seclusion To recognise their rights as full citizens and to work towards realising these rights even though they have never, unlike the elderly been economically productive, and they may never, unlike the disadvantaged child, repay our investment in them by becoming economically productive – is a very deep challenge to a materialist society. Shearer, emphasised that there had been no fundamental questioning of the role of hospitals in providing care that was now recognised to be social and educational and only primarily medical for a small number of people. Meanwhile, vague formulations of 'community care' had left families providing the essential and unpaid services that they always had (Ciçek, 2010).

The CMH promoted an egalitarian ethos, aiming to involve and work alongside learning disabled people, with Shearer advocating therapeutic community-style approaches. As Peter Barham has noted, the egalitarian ideals of 'social therapy' resurfaced and were refashioned in challenges to welfare paternalism and the lack of voice of service users. This is important. The mental hygiene movement had been prominent in the development of social therapies, therapeutic communities and an associated critique of institutionalization as inhibitive of wellbeing. But the psychotherapeutic theories underlying these discriminated against application to mentally deficient people. Only gradually, and predominantly through the work of psychologists, had the movement accepted their relevance. In any case, the movement had considered therapeutic community and social therapy approaches as modernising treatment technologies, not fundamental challenges to hospital-centred care. Nor did the movement aim to develop any potential implications for revising notions of citizenship and its relationship to health and welfare (Cetin, 2004).

Human Condition was mainly oriented towards provisions for people diagnosed mentally ill. Its contention that the voluntary basis of admission could be rendered nugatory by the continued presence of coercive powers, patient passivity or even 'induced collusion' was heavily criticized. Yet this ignores the large majority of the 60 000 people living in mental handicap hospitals who were 'voluntary' patients. As Gostin observed in a footnote, 'The issue of informal admissions for mentally handicapped people is especially well-defined. It is not often that these persons will actively object to their confinement, even though the confinement is not necessarily in their own best interests. The health and welfare of mentally handicapped people was an important part of MIND's rights-based campaigning. Indeed, even while A Human Condition was being published, MIND was working closely with the CMH and beginning to elaborate a framework for community-based services. Over the next few years, in evidence and responses to various government committees, MIND outlined the coordinates of a comprehensive health and welfare scheme that had social integration and citizenship rights at its heart (Cetin, 2004).

The inter-war mental hygiene movement maintained this view, while nevertheless increasingly placing its greatest concern on the mental stability of the wider population. Overstated, the early post-war decades might be summed up in these terms. After the war, mentally deficient people ceased to be considered a threat to society, but few talked about it. In the process, they became citizens, but few noticed. The mental hygiene movement shifted away from vocal concern about the social threat of mental deficiency in the immediate post-war period, but it did not reconsider the fundamental nature and aims of the mental deficiency system. Meanwhile, the inauguration of the Welfare State did little to improve the citizenship status of mentally deficient people, building on the services that had existed before the war and drawing these into the NHS. Additionally, the experiments by psychologists during the 1950s, much cited in the historical literature as significant for changing perceptions towards mentally deficient people, were, in fact, ambiguous regarding citizenship. While confirming their lack of threat to society, the bulk of the research revealed unrecognised abilities to learn and perform routine tasks. This suggested that many people could attain some place in society. But it nevertheless implied that citizenship relied on 'improving' these people so that they could assimilate into the 'lowest' work levels. This left the nature of citizenship rights unexamined. Indeed, the general failure to directly address the relationship of citizenship to mental deficiency helps account for the ambiguities of the 1959 Mental Health Act and subsequent 1960s developments (Boerner, Dütschke, and Schwmmle, 2005).

Conventional histories of post-war learning disability make little reference to psychotherapeutic models; understandably so. Their direct impact has appeared insignificant and more general impact negative. Yet, despite the conceptual bias against people considered mentally deficient, the present analysis shows that, from the 1950s, a discourse of emotionality derived from psychodynamic thinking did gradually have a positive impact.

Under the influence of psychodynamic ideas, mental hygienists developed a general concern with emotional development and wellbeing, emphasizing emotionality as a dynamic relational phenomenon, deteriorated by insensitive relationships and rigid institutional life. This occurred largely through application by professionalizing psychologists, at a time of sustained criticism of the mental deficiency system by the NCCL. Despite the potential ethical issues of the experiment itself, the later Brooklands research expressed a powerful statement of shared humanity and citizenship. Instead of MIND aimed to place the citizenship rights of people with learning disabilities at the heart of comprehensive community-based health and welfare services. The NAMH had been unable to make this shift within the discourse of mental hygiene and the accompanying need to accommodate the professional interests represented within it. But, clearly, MIND's campaigning cannot be reduced to an individualist approach that pitted itself against medicine and psychiatry and founded its stance on considerations of the deprivation of liberty. Some in the psychiatric profession, nevertheless, considered it so.

For instance, in 1984, Kenneth Rawnsley, in a Presidential address to the Royal College of Psychiatrists, maintained that the attempt to remove mental handicap from the remit of the Mental Health Act (which he reduced to a campaign by Mencap) 'boiled down to a reluctance to allow mental handicap to appear in an Act of Parliament, alongside mental illness, since this would somehow stigmatise the former condition'. But this is a caricature. MIND, the CMH and Mencap all sought mental handicap's removal on the basis of research and practical evidence that had accrued since the 1950s. HeatonWard, a former medical adviser to the NAMH, in similar vein, accused the CMH of promoting the idea that mentally handicapped people should be treated 'as though they were, in fact, normal' (Boerner., Dütschke. and Schmmle, 2005).

This is untrue. Along with MIND, the CMH provided detailed analyses of the types of support and care necessary for a community-based service, and set out sophisticated analyses of the internal social organisation necessary within residential provision and the make-up of multi-professional services. Neither can MIND be justifiably considered opposed to psychiatry and medicine. Given that it severely criticised the prevailing medically oriented services for mentally handicapped people and sought a reduced role and authority for psychiatrists within the community, it is understandable that many doctors took this line. But MIND wanted what it considered a more cogent role for psychiatrists within a community service based on citizenship rights.

Notes

References:

1. Abo Tiah B. (2012). The impact of organizational Justice on Organizational Citizenship Behavior in Governmental Ministries Center in Jordan. Islamic University for Economic and Administrative Studies.20 (2): 145-186.

2. Abu Elanain M. (2010). Work Locus of Control and Interactional Justice as Mediators of the Relationship between Openness to Experience and Organizational Citizenship Behavior. Cross Cultural Management: An International Journal. 17(2): 170-192.

3. Acad Matters J (2010), Higher Educ.;5-9. http://www.academic matters.ca/current issue.article.gk?catalog item id= 1234&category=featured articles#. Accessed July 10, 2010.

4. Ahmed A, Fadel K, Ghallab S, Abo El Magd N (2014). Effect of Organizational Justice and Trust on Nurses" Commitment at Assiut University Hospitals. New York Science Journal .7(10): 103-114.

5. Ahmed N, Rasheed A, Jehanzeb K (2012). An exploration of predictors of organizational citizenship behavior and its significant link to employee engagement. International Journal of Business, Humanities and Technology. 2 (4): 99-106.

6. Alabi, G. (2012). Understanding The Relationship Among Leadership Effectiveness, Leader-Member Interactions And Organizational Citizenship Behaviour In Higher Institutions Of Learning In Ghana. Journal of International Education Research. 8(3):263-278.

7. Al-Adel A. (2001). Trajectory Analysis of the Relationship between the Ability to Solve Social Poblems and All the Components of Self-efficacy and the Trend towards Risk. Faculty of Education Journal. Ain Shames University.Part I. 178 (25):121.

8. Albany, NY: State University of New York Press; 2008: 223-239.

9. Allen M, Ogilvie L.(2004), Internationalization of higher education: potentials and pitfalls for nursing education. Int Nurs Rev. 2004;51:73-80.

10. Altınbaş B. (2008). A Practice between Organizational Commitment and Organizational Citizenship Unpublished Master Thesis, Yıldız Technical University, Social Sciences Institute, Administration Department: Pp. 18-27.

11. Altuntaş, S. (2004). Determining the   Attitudes of Nurses Towards Their Job. Istanbul University, Institute of Health Sciences. Istanbul.

12. Altuntaş, S. (2008). The Relationship Between Nurses' Organizational Trust Levels, and Their Personal-Professional Characteristics and Organizational Citizenship Behaviors. Istanbul University, Institute of Health Sciences. Istanbul.

13. Altuntaş, S. and Baykal, U. (2010). Relationship between nurses' organizational trust levels and their organizational citizenship   behaviors. Journal of Nursing Scholarship, 42, 186-194. http://dx.doi.org/10.1111/j.1547-5069.2010.01347.

14. Amikhi F. (2009). The relationship between empowerment and organizational citizenship behavior in educational organizations in Qom. MS Thesis, Tehran University, (College of Qom).

15. Anthea B. and Pauline M. , (1977), Unmet Need: The Case of the Neighbourhood Law Centre (Routledge: London, 1977), 1–4. 54,55,56

16. Association of Universities and Colleges of Canada, (2004), (AUCC). Achieving Canadian excellence in and for the world:

leveraging Canada's higher education and research. Int Aff.; http://www.aucc.ca/ pdf/ english/reports/2004/Cdn Excellence e.pdf. Accessed October 16, 2009.

17. Bekemeier B, Butterfield P. (2005), Unreconciled inconsistencies: a critical review of the concept of social justice in 3 national nursing documents. Adv Nurs Sci.;28(2):152-162.

18. Boerner, S., Dütschke, E. and Schwommle, A. (2005). Doing voluntary extra work? Organizational citizenship behavior in the hospital-A comparison between physicians and nurses. Gesundheitswesen Bundesverband DerArzte Des Offentlichen Gesundheitsdienstes, 67, 770-776. http://dx.doi.org/10.1055/s-2005-858795

19. Boutain D, (2008). Social justice as a framework for undergraduate community health clinical experiences in the United States. Int J Nurs Educ Scholarsh.; 5(1):1-12.

20. Brian Dyson, Liberty in Britain,(1994), : A Diamond Jubilee History of the National Council of Civil Liberties (London: Civil Liberties Trust, 1994), 43.

21. Canadian Nursing Association.(2009) Nursing leadership: do we have a global social responsibility? http://www.cna-aiic.ca/CNA/documents/pdf/publications/ CNA Nursing Leadership Exec Sum e.pdf. Published 2008. Accessed October 16.

22. Cetin, M.. (2004). Organizational Citizenship Behaviors. Nobel Publication Distribution, Ankara.

23. Chavez FS, Peter E, Gastaldo D.(2010), Nurses as global citizens: a global health curriculum at the University of Toronto, Canada. In: Tschudin V, Davis AJ, eds. The Globalization of Nursing. Oxford, UK: Radcliffe Publishing Ltd; 2008:175-186. Copyright© 2010 Lippincott Williams&

Wilkins. Unauthorized reproduction of this article is prohibited.

24. Chen LC, Berlinguer G. (2001), Health equity in a globalizing world. In: Evans T, Whitehead M, Diderichsen F, Bhuiya F, Wirth M, eds. Challenging Inequities in Health: From Ethics to Action.NewYork,NY: Oxford University Press; 2001:35-44.

25. Ciçek, S.S. (2010). Factors Effective on Organizational Citizenship Behaviors: A Model Suggestion. Mustafa Kemal University, Institute of Social Sciences. Hatay.

26. Coss C. Lillian D (1989): Progressive Activist. New York, NY: The City University Press.

27. Dolma, S. (2003). The Effects of Incentives in Displaying Organizational Citizenship Behaviors. Istanbul University, Institute of Social Sciences. Istanbul.

28. Duncan SM, Leipert BD, Mill JE. (1999) "Nurses as health evangelists"? The evolution of public health nursing in Canada, 1918-1939. Adv Nurs Sci.;22(1):4051.

29. Faculty of Nursing University of Toronto, (2010). Nurses as Global Citizens A day in the life of an outpost nurse, Building capacity in Ethiopia, Advancing nursing in Afghanistan, Global health in the curriculum

30. Faculty of Nursing, University of Alberta.(2006) Mission Statement. Edmonton, AB, Canada: University of Alberta; 2006.

31. Geçer, H. (2008). Determination of Organizational Citizenship Behaviors of Nurses in a Hospital. Ankara University, Institute of Health Sciences. Ankara.

32. Güler, B. (2009). The Relation Between Organizational Citizenship Behaviors and Interorganizational Conflict: An

Application in Medical Sector. Marmara University, Institute of Social Sciences. Istanbul.

33. Hanson L. (2009), Global citizenship, global health and the internationalization of the curriculum: a study of transformative potential. J Stud Int Educ. http://jsie. sagepub.com hosted at http://online.sagepub.com. doi: 10.1177/1028315308323207. Published August 22, 2008. Accessed November 1, 2009.

34. Hirschfeld MJ. (2008), Globalisation: good or bad, for whom? In: Tschudin V, Davis AJ, eds. The Globalization of Nursing. Oxford, UK: Radcliffe Publishing Ltd; 2008:12-24.

35. Jahangir, N., Akbar, M.M. and Haq, M. (2004). Organizational citizenship behavior: its nature and antecedents. BRAC University Journal, 1, 75-85.

36. Keeping J, Shapiro D. (2010), Global citizenship: what is it, and what are our ethical obligations as global citizens? Law Now. 2008. http://www.chumirethics foundations.ca/files/pdf/Global%20Citizenship Law Now jk ds0708.pdf. Accessed July 10, 2010.

37. Kelley MA, Connor A, Kun KE, Salmon M, (2008). Social responsibility: conceptualization and embodiment in a school of nursing. *Int J Nurs Educ Scholarsh.*;5(1):1-16.

38. Knight J.(2008), The internationalization of higher education: are we on the right track

39. Koberg, C.S., Wayne. B.R., Goodman, E., Boss, A. and Monsen, E. (2005). Empirical

40. evidence of organizational citizenship behavior from the health care industry. International Journal of Public Administration, 28, 417-436. http://dx.doi.org/10.1081/PAD-200055199

41. Kөse, S., Kartal, B. and Kayalı, N. (2003). A study on the relationship between factors related to organizational citizenship behavior and attitude. Erciyes University Journal of Economics and Administrative Sciences Faculty, 20, 1-19.

42. Maria Mayer-Kelly and Michael D. Kandiah (2016), 'The Poor Get Poorer Under Labour: the Validity and Effects of CPAG's Campaign in 1970', ICBH Witness Seminar Programme, seminar held 18 October 2000, (Institute of Contemporary British History, 2003, http://www.icbh.ac.uk/icbh/witness/cpag/,

43. Mayo K. (1996), Social responsibility in nursing education. J Holist Nurs. 1996;14(1):24-43.

44. NAMH, (1971), MIND Manifesto, republished in MIND and Mental Health, 1 3–6: 4. SA/MIN/B/80/27/28.

45. NAMH, (2013), Annual Report, 1971–2, 12. SA/MIN/B/80/7/3. Paul Williams, 'The Roots of a True Campaigning Movement', Community Living, 26, 3 (2013), 17. NAMH Council of Management, 18 June 1971. SA/MIN/A/3/2.

46. NAMH, Annual Report 1965–6, 5; Council of Management Minutes, paper on 'National Trends and The Mental Health Services' 21 November 1967. SA/MIN/A/3/1.

47. Organ, D.W., Podsakoff, P.M. and MacKenzie, S.B. (2006). Organizational Citizenship Behavior: Its Nature, Antecedents, and Consequences. SAGE Publications, Inc, California.

48. Park, J., Yun, E. and Han, S. (2009). Factors influencing nurses' organizational citizenship behavior. Journal of Korean Academician Nursing, 39, 499-507. http://dx.doi.org/10.4040/jkan.2009.39.4.499

49. Pike G.(2008), Reconstructing the legend: education for global citizenship. In: Abdi AA, Shultz L, eds. Educating for Human Rights and Global Citizenship.

50. Regan et al. (1973), 1.Hilary Rose, 'Up Against the Welfare State: The Claimant Unions', Socialist Register, 10 (1973), 179–203.

51. Reimer-Kirkham S, Browne AJ(2006). Toward a critical theoretical interpretation of social justice discourses in nursing. Adv Nurs Sci.;29(4):324-339.

52. Sabuncuoğlu, Z. and Tüz, M. (2005). Organizational Psychology. Alfa Aktüel Publication Distribution Co. Ltd. Bursa.

53. Schnake, M.E. and Dumler, M.P. (2003). Levels of measurement and analysis issues in organizational citizenship behavior research. Journal of Occupational and Organizational Psychology, 76, 283- 301.

54. Smith SM. (2002) Nursing as a social responsibility: implications for democracy from the life perspective of Lavinia Lloyd Dock (1858-1956) (doctoral disserta-tion, Louisiana State University and Agricultural & Mechanical College, 2000). Dissertation Abstracts Int. 2002;63(11):3846A (AAT 3069733).

55. Stable URL: http://www.jstor.org/stable/3406147Accessed: 21-08-2017 08:23 UTCStop.

56. Stanley D.2017 Nursing and Citizenship, The American Journal of Nursing, Vol. 16, No. 1 (Oct., 1915), pp. 15-24 Published by: Lippincott Williams & Wilkins

57. Tarshall,(1950), Citizenship and Social Class and Other Essays (Cambridge: Cambridge University Press.

58. Tamara Goriely,1999) 'Making the welfare state work: changing conceptions of legal remedies within the British welfare state', in Francis Regan, Alan Paterson and Tamara Goriely (eds), The Transformation of Legal Aid: Comparative and Historical Studies (Oxford: Oxford University Press,), 89–112.

59. Technology, Bioethics Section; 2000. SHS/EST/CIB 10-11/1. http://unesdoc.unesco.org/images/0018/ 001878/187899E.pdf. Accessed July 10, 2010.

60. Torres CA, Rhoads RA. (2006), Introduction: globalization and higher education in the Americas. In: Rhoads RA, Torres CA, eds. The University, State and Market: The Political Economy of Globalization in the Americas. Stanford, CA: Stanford University Press; 2006:3-38.

61. United Nations Educational,(2010), Scientific and Cultural Organization. Report of the International Bioethics Committee of UNESCO (IBC) on Social Responsibility and Health. Paris, France: Social and Human Sciences Sector, Division of Ethics of Science

62. University of Alberta, (2008). Connecting with the World: A Plan for International Engagement. Edmonton, AB, Canada: University of Alberta; 2008.

63. Zee, E. (2009). Organizational citizenship behavior. Keser, A., Yılmaz, G. and Yürür, S. (Eds.). Behavior in Working Life. Umuttepe Publication, Kocaeli.